# Remotely Piloted Aircraft
## and
## War in the Public Relations Domain

The well-intentioned author of the article "The Killing Machines," which appeared in the *Atlantic* last year, offers a lengthy description of a Hellfire missile strike by a remotely piloted aircraft (RPA). The story's protagonist, a "19-year-old American soldier" who entered Air Force basic military training straight out of high school, became an MQ-1 Predator crew member upon graduation. Reportedly, on his very first mission at the controls, the "young pilot" observed a troops-in-contact situation on the ground. The "colonel, watching over his shoulder, said, 'They're pinned down pretty good. They're gonna be screwed if you don't do something.'"[1] The narrative goes on to describe the Hellfire missile strike and the psychological effect it had on the Airman.

To a sophisticated military audience, the factual inconsistencies in this account are apparent. Air Force RPAs are crewed by Airmen, not Soldiers. The 19-year-old Airman (an enlisted rank) cannot be an Air Force pilot (an officer rating). The article claims that during his first time at the controls, this Airman finds himself on a combat mission in-theater. In reality, he would have become familiar with the controls at initial qualification training, prior to arriving at his first combat squadron. Furthermore, when colonels speak to Airmen about life-and-death combat decisions, they tend to do so in terms of direct orders rather than leading suggestions. How can Mark Bowden, notable historian and author of such well-received books as *Black Hawk Down*, commit such factual errors? The answer is simple. Information about Air Force RPA operations is rarely available—and when it is, it usually proves unreliable. This article contends that because an information vacuum exists with respect to US RPA operations, well-meaning people cannot gain adequate knowledge to develop and share an informed opinion on the most important RPA questions. It calls this dearth of information "the epistemic problem."

To disprove a deductive argument, one must (a) disprove one or more premises, (b) identify an ambiguous definition, or (c) demonstrate a logical fallacy in the argument.[2] Many of the RPA articles, opinions, and interviews produced over the last decade are either based on false premises (option a) or employ a logical fallacy of analogy (option c); therefore, many of their conclusions are invalid. This article does not attempt to show that most of the writing on RPAs over the last decade contains fallacies of some kind. Rather, it recognizes the ease with which sincere people can commit such errors as a result of the epistemic problem inherent in any discussion of RPA operations.

The argument, then, begins by asserting that such a problem exists and suggesting that it has three causes. First, enemy forces (here referring specifically to al-Qaeda and the Taliban) have an effective public relations (PR) campaign against RPAs. Second, the United States conducts an ineffective PR campaign in support of RPAs. Third, RPA operations are necessarily concealed by security classifications and national security precautions. The article expounds upon the significance of these causes and provides evidence for them—evidence that will demonstrate not only the three causes but also the reality of the epistemic problem. Its conclusion offers two ways that individuals can mitigate the dilemma and one means by which the US government can rectify it.

## Enemy Forces and Public Relations

The term *propaganda* is omitted here because it is controversial and because, even assuming a universally agreed-upon definition, identifying its individual instantiations would prove difficult. For example, one definition holds that "propaganda is biased information designed to shape public opinion and behavior."[3] Another tries to circumvent a negative connotation by distinguishing weak from strong propaganda, describing the former as "persuasion in the interests of the message sender, based on selected facts and emotions."[4] Even if information meets the criteria established by these definitions, it does not necessarily warrant the negative connotation often intended by the term *propaganda*. Although governments and terrorist organizations may engage in it, the term remains unhelpful. The fact that information is biased does not make it false, and the fact that information intends to shape public opinion and action does not make it underhanded or deceitful. This article concerns itself with the genus of information, within which propaganda is a species, and therefore addresses all information—biased and unbiased, true and false—designed to shape public opinion and action.

*PR*, then, is a better term because it sidesteps the potentially pejorative connotation of the other term and associates a particular set of information with a particular organization. Such an entity with a PR arm, committee, or campaign carefully crafts its message to achieve certain aims. Although some information publicized by al-Qaeda and by an American news network may overlap (regarding a specific RPA strike, for example), that news network is not *participating* in al-Qaeda's PR campaign. (Indeed, if it relies upon information published by that militant group, then the news agency may be a *victim* of the campaign instead of a participant.)

The effects of al-Qaeda's and the Taliban's PR efforts are noticeable and far reaching. In an independent international poll conducted in 2012, the vast majority of respondents were strongly opposed to the US RPA campaign.[5] In his article "A Progressive Defense of Drones," Yale Human Rights Fellow Kiel Brennan-Marquez notes that "as a liberal, I'm against drones essentially by reflex. . . . Unlike traditional warfare, when the loss of life on the other side is presumptively *acceptable*, . . . in the case of drone strikes, the loss of lives on the other side is presumptively *unacceptable*"[6] (emphasis in original). Why these presumptions? Why does the world seem preconditioned against RPAs? What is it about their operations that makes them seem inherently antiliberal or presumptively unacceptable? Some of these seemingly intuitive responses may actually be conditioned, based upon the public's exposures to RPA operations. Such exposure is controlled—or at least influenced—by intentional PR campaigns. The following discussion presents three different models that explain the level of influence and intentionality of anti-RPA PR campaigns, ordering them from the most benign (which assigns a passive role to the PR campaign) to the most severe (which assigns an active role to an enemy PR campaign).

In their ethical assessment of targeted killing, Eric Patterson and Teresa Casale contend that "while contemporary targeted killing is useful for striking terrorists in dangerous places, it will be covered by foreign media like al Jazeera in ways unflattering to the U.S."[7] In this view, reporting biases are based on inherent cultural relationships between the news agencies involved and the victims or proponents of RPA strikes. Therefore this "cultural relationship" perspective assigns no PR intentionality to the news agency. That is, these authors do not suggest that al Jazeera is participating in the enemy's PR campaign, or any other, against RPAs. Any biased reporting is simply based upon the nature of things as they are. So American news agencies are just as likely to be slanted toward US interests. Thus, such agencies will necessarily produce biased information without intentionally engaging in a PR campaign.

Other individuals, though, take a stronger view. Regarding many civilian casualty reports, Georgetown professor Daniel Byman claims that "their numbers are frequently doctored by the Pakistani government or by militant groups. After a strike in Pakistan, militants often cordon off the area, remove their dead, and admit only local reporters sympathetic to their cause or decide on a body count themselves. The U.S. media often then draw on such faulty reporting to give the illusion of having used multiple sources."[8] This "controlled information" view is stronger than that of the "cultural relationship" insofar as it does suggest that some groups have a PR agenda. These groups, however, are purportedly not enemy forces but third-party, anti–United States groups—in this case, non-US governments and militant groups.

Prof. Audrey Kurth Cronin of Oxford University and George Mason University takes the strongest position: "Al Qaeda uses the strikes that result in civilian deaths, and even those that don't, to frame Americans as immoral bullies who care less about ordinary people than al Qaeda does." (She notes that this PR strategy is effective in spite of the fact that US RPA strikes avoid civilians about 86 percent of the time and that al-Qaeda purposefully targets them.)[9] This is the "enemy PR campaign" view—the most plausible of the three—which asserts that intelligent people within the enemy's organizational structure intentionally affect information streams so that passive recipients (global populations) will condemn the United States' use of RPAs.

Some people may be tempted to doubt an active al-Qaeda or Taliban PR campaign on the grounds that such a decentralized organization or set of terrorist cells probably does not have the strategic capability to affect information to the degree required to sway global opinion. In the face of such doubt, though, one must recall that international terrorism is by its nature a PR endeavor. US Army Field Manual 7-98, *Operations in a Low Intensity Conflict*, cautions that a terrorist organization's acts of violence "draw the attention of the people, the government, and the world to . . . [its] cause" and that "the media plays a crucial part in this strategy."[10] Megan Smith and James Igoe Walsh note that al-Qaeda is among those terrorist groups that "calculate the consequences of their actions not only in the number of lives lost or the economic and social damage inflicted, but in the amount of media attention they are able to garner."[11] In what has been called "mass-mediated terrorism," organizations such as al-Qaeda not only spin media-produced coverage of their activities but also produce their own coverage.[12] In this way, al-Qaeda can generate a message and shape, control, and distribute it to maximum effect. In fact, it is so sophisticated in this domain that it has a designated PR branch called as-Sahab ("the Cloud") Media.[13]

RPA operations are vulnerable to as-Sahab's PR machine in two ways. First, RPA strikes "provide as-Sahab with incidences of United States behavior that can be painted as cruel, brutal, and capricious to a mass audience, further legitimizing the political stances of al Qaeda."[14] Second, as-Sahab can attack the nature of RPA warfare without reference to particular strikes. Indeed, it has released "numerous public statements asserting that the United States exploits its unfair advantage in technology and that the use of unarmed drones is cowardly."[15]

Like as-Sahab, the Taliban can affect public opinion regarding the use of remote weapons, though perhaps not with the same level of sophistication. Target audiences particularly vulnerable to such influence include the local populations of Afghanistan and northwest Pakistan. In the past, in the absence of US government commentary on reported RPA strikes, the Taliban have taken full advantage of the silence, convincing the local populace that insurgent effects were actually caused by the coalition. They have been so convincing that in 2009, the Taliban successfully convinced Kandahar City residents that an explosion caused by a Taliban bomb years earlier was actually the result of a US air strike.[16]

In addition to the enemy's top-down PR directives, digital interconnectivity and social media have allowed for PR efforts orchestrated at the middle-management level. Zachary Adam Chesser, a 20-year-old American, earned notoriety among jihadists when he e-published an insider's guide to defeating the United States in the PR domain. He urged his adherents to "publish statistics of how many Muslim civilians have been killed by the Americans, using the highest credible estimates. . . . Anytime an American does something wrong, emphasize it . . . [and] anytime the United States does anything that can be perceived as a success in its war against al Qaeda, bury it."[17] Al-Qaeda and the Taliban have recognized and exploited the PR domain for a kind of information superiority and have subsequently poisoned many people against RPA operations—both within the United States and abroad. Thus far, we have learned that the enemy conducts an active PR campaign and that one of its highest priorities is tarnishing the world's opinion of RPAs—perhaps the most capable weapons against that adversary.[18]

## The United States' Response

America has met the enemy's PR effectiveness with its own PR failures. Misconceptions about RPA operations have been widespread and continue to proliferate. Take for example the "video game problem." Bowden says that killing from 3,000 miles away is "like a video game; it's like Call of Duty."[19] Professor Brennan-Marquez claims that the "numbness that results from using machines rather than soldiers to carry out our dirty work" produces "the nightmarish image of an 18-year-old drone operator basically playing video games from the detached safety of a Nevada bunker."[20] Moreover, the subtitle of an article by Michael Brooks, a science journalist and holder of a PhD in quantum physics, in the *New Statesman* reads, "Can You Play a Video Game? Then You Can Fly a Drone."[21]

This video game argument employs a logical fallacy called "a failure to recognize distinctions" by D. A. Carson and a "faulty analogy" by Norman Geisler and Ronald Brooks.[22] This method of refuting an argument reflects option C (above), demonstrating that the conclusion does not follow from the premises. Such a faulty analogy occurs by assuming that when two things are similar in one way they will be similar in another way.[23] Proponents of the video game hypothesis claim that flying an RPA is *like* playing a video game—and they may be quite right. But the fact that the two are alike in one way does not mean that they are alike in all ways.

The video game pejorative is rhetorical in nature, and its negative connotation is apparent: RPA pilots must see as little correspondence between their activities and reality as do the video gamers. The pilots must not take their job seriously, just as people who play a video game are not serious, and they look at killing real people in the same way that the video gamers perceive killing computer-generated people. In this way, there exists a necessarily cognitive and emotional distance, as well as a disinterested detachment, from the death that the pilots administer—or so the claimants would have us believe.

Expressed in these terms, the video game hypothesis obviously becomes inadequate—and so it is. Nevertheless, it is pervasive enough to deserve attention. First, one must discover similar elements between the video game and the RPA. Although proponents of the hypothesis should do this themselves, let us consider one possibility. The RPA pilot, like the video gamer, sits in a dark, air-conditioned room with multiple video monitors, a headset, and a microphone, having no exposure to the physiological pressures of manned flight. If these are, in fact, the elements shared by the two activities, then two responses to the video game hypothesis emerge. The first entails identifying the fallacy and asserting the fact that the existence of similar elements does not imply the similarity of all elements. This might prove a weak response, however, in that even though it demonstrates that the important elements (dissociation with reality, etc.) are not *necessarily* similar, it does not demonstrate that they are *dissimilar*.

A second, stronger response involves identifying another activity that resembles the video game in the same way that the RPA resembles it but that at the same time is dissimilar with respect to the important issues (dissociation with reality, etc.). Two obvious examples come to mind. The first is a radar-approach controller at a busy airport, such as Boston's Logan International. This individual sits in a dark room, looks at several video monitors, and wears a headset and a microphone, having no exposure to the physiological pressures of flight. He or she, though, may control multiple airliners, each carrying hundreds of people in instrument meteorological conditions (i.e., the aircraft are in the weather and rely upon instruments, navigational aids, and the controller's instructions). In this case, the mechanism and aesthetics of the controller's job are strikingly similar to those of the video gamer and RPA pilot, yet the controller appears to face no "video game" critique in the popular press or scholarly literature. Furthermore, one could argue that if the controller were to dissociate his or her activity from reality, the results would prove even more catastrophic than those that would follow if the RPA pilot were to do the same. On the one hand, the MQ-1's two 100-pound Hellfire missiles give the pilot only limited destructive power.[24] On the other hand, if the controller runs two Boeing 737s together (a relatively small airplane at Boston Logan), then more than 250 people face a high probability of death. No RPA strike has generated that many casualties.

The combined force air component commander (CFACC), who has operational control of RPA missions in the area of responsibility, offers an additional counterexample.[25] That general officer also sits in a dark, air-conditioned room with a number of video monitors. Again, demonstrating greater destructive capability than the RPA pilot, this officer is responsible for numerous missions, objectives, and air assets. Are the proponents of the video game hypothesis prepared to accuse the CFACC of dissociating his activity from reality? If so, there should be as many articles published about the dangers of the Defense Department's entire command and control architecture as there are warnings against video-game-like weapons systems.

The video game problem offers the best example of a PR failure that the US government could rectify with a better PR campaign. People think that flying an RPA is like playing a video game, in part, because of their limited exposure to the operations of that platform. After all, they see only the video-game-like apparatus of a dark room, video monitors, a headset, and a microphone—but no flight physiology (see the Air Force's own television advertisements).[26] It should come as no surprise that they extend the analogy between RPA and video games beyond its legitimate reach. This problem, though, is not the only one faced by the US PR campaign.

A number of other false claims about the United States' RPA capabilities can be reduced to doubts about discrimination. The just war tradition and, indeed, America's own law of armed conflict require that the United States (and any belligerent, in the case of just war) discriminate between combatant and noncombatant.[27] Professor Brennan-Marquez asserts that "death, visited from the skies, isn't precise."[28] The advocacy group Anti-War Committee claims that "the physical distance between the drone and its shooter makes a lack of precision unavoidable."[29] Political scientist Michael Gross suggests that in a conflict with a nonstate actor such as al-Qaeda, militaries (including the US armed forces) most likely will assume that individuals in civilian clothes are combatants until otherwise demonstrated.[30]

The United States' failure to disclose information about its own use of RPAs has produced an additional noteworthy consequence. As was the case with nuclear weapons, America has been among the first to attain this kind of remote capability.[31] That said, Dr. Micah Zenko, a fellow at the Council on Foreign Relations, correctly points out that "over the next decade, the U.S. near-monopoly on drone strikes will erode as more countries develop and hone this capability."[32] Just as the United States was in a position to craft the global standard for nuclear weapons practices, so can it help establish international norms for the acceptable use of remote weapons.[33] Despite its failure to disclose information critical to this cause, the US government has acknowledged the inevitable proliferation of remote weapons. According to the Obama administration, "If we want other nations to use these technologies responsibly, we must use them responsibly."[34] Given the heavy veil covering the RPA program, however, the international community cannot determine the United States' degree of responsibility. The disclosures recommended in this article would not only allow for but also foster international discussion on the acceptable use of remote weapons—a discussion that some would argue is imperative.[35]

The US government could address all of these issues by making two important disclosures, neither of which would violate national security requirements. First, it could publicize the already unclassified capabilities of the RPA weapons systems.[36] Assertions that RPA strikes are by nature indiscriminate are false. Though conditioned by an effective al-Qaeda PR campaign to believe otherwise, people who have done the research have found this to be the case. As Avery Plaw, a political scientist at the University of Massachusetts, observes, "The drone program compares favorably with similar operations and contemporary armed conflict more generally."[37] The International Committee of the Red Cross found that throughout armed conflicts of the twentieth century, 10 civilians were killed for every combatant.[38] Because the accounts vary so widely, a precise civilian casualty rate for RPA strikes is impossible to determine. Nevertheless, it is certainly less than .5 civilians per combatant and may be as low as .08 (20 to 125 times better than the historical standard set by twentieth-century conflicts).[39]

Second, the United States could publicize elements from its own internal rules of engagement for distinguishing civilians from enemy combatants.[40] In January 2012, leaders of the Afghanistan International Security Assistance Force (ISAF) met to discuss methods of eliminating civilian casualties in Afghanistan. Lt Gen Adrian Bradshaw, the ISAF's deputy commander at the time, told attendees that "eliminating Afghan civilian casualties is a high priority" and even a "moral obligation."[41] Countering Professor Gross's assumption above, the ISAF's priority suggests that the US military presumes *civilian* status until otherwise demonstrated and not the other way around. Further disclosures like this one, not of details but of priorities and general practices, would help assure the world's population that the United States takes the just war tradition's requirement of discrimination quite seriously.

## Classification and Secrecy

As Professor Rosa Brooks rightly observes, the United States' use of RPAs is shrouded in secrecy.[42] This is true, as she suggests, not only for targeted killings but also for close air support operations in Afghanistan. This article distinguishes the poor PR campaign from issues of classification because the requirement to win the PR war does not supersede the one to keep classified material out of the hands of the enemy and, therefore, out of the hands of the public.

The appropriate way to view the epistemic problem in this context calls for recognizing two constants and one variable. The enemy's PR campaign resembles a constant in that it lies outside the US government's control. After all, in the PR domain, as in all the others, the enemy gets a vote. One should also consider classification a constant. Reasons exist for classifying information and for winning in the PR domain, but those reasons are independent of each other. One cannot expect the motivations for an effective PR campaign to outweigh those for classification; consequently, one should not expect to change the way the US military makes classification determinations for the sake of an effective PR campaign. In the context of such an effort, then, classification should be considered a constant. The one variable that America does control in the PR domain is its own PR campaign, discussed in the previous section.

## Conclusion: The Way Ahead

The individual cannot solve the epistemic problem. One can, however, make two significant interpretive moves in reading and writing about RPA operations. First, one ought to know that the problem exists and interpret the information appropriately.

As long as the US government maintains radio silence on its RPA program, responsible readers must recognize that they are receiving only one side of a necessarily polarized story. Once readers realize that an enemy with a sophisticated and well-practiced PR machine at its disposal is engaged in information warfare, using the media as an instrument, they should view these reports cautiously rather than dogmatically. Such is the nature of the epistemic problem. Drawing merely upon news reporting, we simply cannot know exactly what happened in the cockpit or on the ground in a particular RPA strike. So what can we know? After we become aware of the epistemic problem, our interpretation of available data should concentrate on big questions rather than little ones.

The epistemic problem may result in insufficient information to determine whether RPAs are more or less discriminate than their traditionally manned sister platforms—but this is a little question. A big question is whether RPA technology changes the nature of discrimination. The evidence suggests that it does not. The epistemic problem may produce misunderstandings about how flying an RPA is like playing a video game—but this is a little question. A big question is whether the digital apparatus constitutes a sufficient condition for the dissociation between activity and reality. Even one case of post-traumatic stress disorder in an RPA crew member would indicate that it does not.[43] Whether a single RPA crew errs on a single RPA mission is (by comparison only) a little question. A big question is whether the RPA weapons systems in question provide a means for the crew to distinguish reliably between friend and foe.

A second duty in light of the epistemic problem is extraneous to the preceding argument but nevertheless necessary. Anyone speaking or writing about these issues has an obligation to make clear to the audience precisely the type of RPA, kind of mission, three-letter government agency, or area of responsibility under discussion. For example, some have argued that RPAs are unethical in that their use entails no risk.[44] Sometimes, though, risk is in fact present (e.g., when RPAs conduct close air support missions, the enemy may impose great risk on ground forces). These arguments, then, fail to distinguish between conflicts (like Afghanistan) that use RPAs to protect ground troops and notional conflicts that would use *only* RPAs to pursue military ends.[45] Similarly, some have argued that the use of RPAs makes the decision to go to war too easy, again based on absence of risk.[46] This argument also assumes an RPA-only war (a decision to go to war and use RPAs to support ground troops will still come at a heavy cost). The ensuing conclusions may prove valid for some future events, but theorists err in applying them to RPA operations in Afghanistan. Appropriately distinguishing between different uses of RPAs will limit confusion and mitigate the epistemic problem.

The US government can take a significant step toward solving the RPA epistemic problem—and such institutional action would be far more efficacious than that of individuals. As mentioned above, the US government, designating the Department of the Air Force as the lead agency, should conduct an active, international PR campaign in which it publicizes true information, showcasing the discriminatory capability of RPA weapons systems as well as internal safeguards (such as rules of engagement) against haphazard targeting. To this point, the world has heard only one side of a two-sided discussion and, unsurprisingly, has succumbed to it. Intelligent, well-intentioned people should have the opportunity to hear both sides so that they can develop an informed opinion.

All is not lost. An epistemic problem exists, but a meaningful conversation can commence nevertheless. Awareness of the problem should influence one's thoughts and actions. Additionally, one should not submit to an omniscient technocracy, trusting that those privy to the secrets must know best and that, therefore, the individual need not know anything about it at all. On the contrary, to the extent that national security can be safeguarded, this article has argued that the federal government should not simply disclose but *publicize* much of its RPA program that remains in the dark. The battle for hearts and minds with respect to RPAs is being waged in the PR domain. Today, the enemy is winning. ✪

## Notes

1. Mark Bowden, "The Killing Machines: How to Think about Drones," *Atlantic*, 14 August 2013, http://www.theatlantic.com/magazine/archive/2013/09/the-killing-machines-how-to-think-about-drones/309434/. The unmanned aircraft name game is ongoing. The US Air Force uses the term *remotely piloted aircraft* (*RPA*) whereas the other services use *unmanned air systems* (*UAS*). Nonmilitary sources typically refer to *drones*. Each term refers to the same group of systems. More importantly, this article refers to armed systems (most notably, Air Force MQ-1s and MQ-9s) operated by military personnel located in the continental United States. The article preserves the terms as they appear in quoted texts.

2. Peter Kreeft, *Socratic Logic: A Logic Text Using Socratic Method, Platonic Questions, and Aristotelian Principles*, ed. 3.1 (South Bend, IN: St. Augustine's Press, 2010), 26.

3. "What Is Propaganda?," United States Holocaust Memorial Museum, accessed 7 December 2013, http://www.ushmm.org/propaganda/resources/.

4. Johanna Fawkesa and Kevin Moloney, "Does the European Union (EU) Need a Propaganda Watchdog like the US Institute of Propaganda Analysis to Strengthen Its Democratic Civil Society and Free Markets?," *Public Relations Review* 34, issue 3 (September 2008): 209.

5. The poll, though, used the term *drone*. Audrey Kurth Cronin, "Why Drones Fail: When Tactics Drive Strategy," *Foreign Affairs* 92, no. 4 (July/August 2013): 49.

6. Kiel Brennan-Marquez, "A Progressive Defense of Drones," *Salon*, 24 May 2013, http://www.salon.com/2013/05/24/a_progressive_defense_of_drones/.

7. Eric Patterson and Teresa Casale, "Targeting Terror: The Ethical and Practical Implications of Targeted Killing," *International Journal of Intelligence and Counterintelligence* 18, no. 4 (21 August 2005): 647.

8. Daniel Byman, "Why Drones Work: The Case for Washington's Weapon of Choice," *Foreign Affairs* 92, no. 4 (July/August 2013): 37.

9. Cronin, "Why Drones Fail," 47.

10. Field Manual 7-98, *Operations in a Low-Intensity Conflict*, 19 October 1992, 3-1, http://www.bits.de/NRANEU/others/amd-us-archive/fm7-98(92).pdf.

11. Megan Smith and James Igoe Walsh, "Do Drone Strikes Degrade Al Qaeda? Evidence from Propaganda Output," *Terrorism and Political Violence* 25, no. 2 (10 January 2013): 314, http://www.jamesigoewalsh.com/tpv.pdf.

12. Ibid.

13. Ibid.

14. Ibid., 316.

15. Ibid.

16. Sarah Holewinski, "Do Less Harm: Protecting and Compensating Civilians in War," *Foreign Affairs* 92, no. 1 (January/February 2013): 17.

17. Jarret Brachman, "Watching the Watchers: Al Qaeda's Bold New Strategy Is All about Using Our Own Words and Actions against Us. And It's Working," *Foreign Policy*, 12 October 2010, http://www.foreignpolicy.com/articles/2010/10/11/watching_the_watchers. These quotations are from Brachman's paraphrase and are not necessarily Chesser's exact words.

18. Andrew Callam, "Drone Wars: Armed Unmanned Aerial Vehicles," *International Affairs Review* 18, no. 3 (Winter 2010), http://www.iar-gwu.org/node/144.

19. Bowden, "Killing Machines."

20. Brennan-Marquez, "Progressive Defense of Drones."

21. Michael Brooks, "Eyes in the Sky: Can You Play a Video Game? Then You Can Fly a Drone," *New Statesman* 141, issue 5110 (18 June 2012): 27–29.

22. D. A. Carson, *Exegetical Fallacies*, 2nd ed. (Grand Rapids, MI: Baker Academic, 1996), 92; and Norman L. Geisler and Ronald M. Brooks, *Come, Let Us Reason: An Introduction to Logical Thinking* (Grand Rapids, MI: Baker Books, 2004), 109.

23. Kreeft, *Socratic Logic,* 102.

24. "MQ-1B Predator," fact sheet, US Air Force, 20 July 2010, http://www.af.mil/AboutUs /FactSheets/Display/tabid/224/Article/104469/mq-1b-predator.aspx; and "AGM-114R Multi-Purpose HELLFIRE II: Effective against 21st Century Threats," Lockheed Martin, 2011, 2, http://www.lockheedmartin.com/content/dam/lockheed/data/mfc/pc/hellfire-ii -missile/mfc-hellfire-ii-pc.pdf.

25. Or the director of the combined air operations center to whom targeting authority is delegated. Changing this notional example to a joint force air component commander would not affect its application to my argument. See Joint Publication 3-30, *Command and Control of Joint Air Operations*, 10 February 2014, E-1, http://www.dtic.mil/doctrine/new_pubs /jp3_30.pdf; and Curtis E. LeMay Center for Doctrine Development and Education, "Annex 3-0, Operations and Planning," 9 November 2012, 2n2, https://doctrine.af.mil/download .jsp?filename=3-0-Annex-OPERATIONS-PLANNING.pdf.

26. "Remotely Piloted Aircraft," Videos & More, US Air Force, accessed 3 July 2014, http://www.airforce.com/games-and-extras/videos/?cat=12&slug=rpa&catName=AF%20 in%20Action#.

27. Brian Orend, "War," in *Stanford Encyclopedia of Philosophy*, ed. Edward N. Zalta (Stanford, CA: Metaphysics Research Lab, Center for the Study of Language and Information, Fall 2008), http://plato.stanford.edu/archives/fall2008/entries/war/; and US Army Judge Advocate General's Legal Center and School, *Law of Armed Conflict Deskbook* (Charlottesville, VA: US Army Judge Advocate General's Legal Center and School, International and Operational

Law Department, 2013), 13, http://www.loc.gov/rr/frd/Military_Law/pdf/LOAC-Desk book-2013.pdf.

28. Brennan-Marquez, "Progressive Defense of Drones."

29. Harrison Schmidt, Jennie Eisert, and Meredith Aby, "Stop Drone Warfare!," Anti-War Committee, 20 June 2012, http://antiwarcommittee.org/2012/06/20/stop-drone-warfare/.

30. Michael L. Gross, "Assassination and Targeted Killing: Law Enforcement, Execution or Self Defence?," *Journal of Applied Philosophy* 23, no. 3 (August 2006): 329.

31. "To date, only the United States, Israel, and the United Kingdom are believed to have used armed drones." Sarah Kreps and Micah Zenko, "The Next Drone Wars: Preparing for Proliferation," *Foreign Affairs* 93, no. 2 (March/April 2014), http://www.foreignaffairs.com /articles/140746/sarah-kreps-and-micah-zenko/the-next-drone-wars.

32. Micah Zenko, *Reforming U.S. Drone Strike Policies*, Council Special Report no. 65 (New York: Council on Foreign Relations, Center for Preventive Action, January 2013), 4.

33. Kreps and Zenko "Next Drone Wars."

34. Zenko, *Reforming U.S. Drone Strike Policies*, 5.

35. Kreps and Zenko "Next Drone Wars."

36. I distinguish here between the terms *declassify* and *publicize*. Unclassified information is available but not necessarily actively publicized by the US government. I am not advocating the declassification of information but merely publicizing to passive recipients the information already available to someone actively searching for it.

37. Scott Shane, "The Moral Case for Drones," *New York Times*, 14 July 2012, http://www .nytimes.com/2012/07/15/sunday-review/the-moral-case-for-drones.html?_r = 0.

38. Rosa Brooks, "What's *Not* Wrong with Drones: The Wildly Overblown Case against Remote-Controlled War," *Foreign Policy*, 5 September 2012, http://www.foreignpolicy.com /articles/2012/09/05/whats_not_wrong_with_drones#sthash.eCd2QZZj.dpbs.

39. Ibid.

40. Brennan-Marquez, "Progressive Defense of Drones."

41. Lt Col David Olson, "ISAF Conducts Aviation Civilian Casualty Conference," Afghanistan International Security Assistance Force, 19 January 2012, http://www.isaf.nato.int /article/news/isaf-conducts-aviation-civilian-casualty-conference.html.

42. Rosa Brooks, "The War Professor: Can Obama Finally Make the Legal Case for His War on Terror?," *Foreign Policy*, 23 May 2013, http://www.foreignpolicy.com /articles/2013/05/22/the_war_professor#sthash.3OQBfgms.dpbs.

43. This argument hinges on the presupposition that video games do not result in cases of post-traumatic stress disorder—a presupposition upon which I am willing to rest.

44. Paul W. Kahn, "The Paradox of Riskless Warfare," *Philosophy and Public Policy Quarterly* 22, no. 3 (Summer 2002): 2–8, http://digitalcommons.law.yale.edu/cgi/viewcontent .cgi?article = 1325&context = fss_papers; and Suzy Killmister, "Remote Weaponry: The Ethical Implications," *Journal of Applied Philosophy* 25, no. 2 (May 2008): 121–33.

45. Someone may respond that reported RPA strikes in Yemen, Pakistan, and Somalia are representative of this case (an RPA-only conflict). If these strikes are carried out under the Authorization for the Use of Military Force, signed by Congress in 2001, in legal terms they are against al-Qaeda and affiliates and are therefore grounded in defense of US persons at home and abroad. See Brooks, "War Professor."

46. John Kaag and Sarah Kreps call this the moral hazard. See their article "The Moral Hazard of Drones," *New York Times*, 22 July 2012, http://opinionator.blogs.nytimes .com/2012/07/22/the-moral-hazard-of-drones/. See a similar argument formulated as "The Real Problem with Asymmetry," in Jai C. Galliott, "Uninhabited Aerial Vehicles and the Asymmetry Objection: A Response to Strawser," *Journal of Military Ethics* 11, no. 1 (January 2012): 62.